Scientific Freedom and Responsibility

A Report of the AAAS Committee on Scientific Freedom and Responsibility

Prepared for the Committee by
JOHN T. EDSALL

American Association for the Advancement of Science
Washington, D.C. 1975

Responsibility for statements of fact and expressions of opinion contained in this report rests with the AAAS Committee on Scientific Freedom and Responsibility. The AAAS Board of Directors, in accordance with Association policy and without passing judgment on the views expressed, has approved its publication as a contribution to the discussion of an important issue.

This report was prepared for the Committee by John T. Edsall.

Printed in the United States of America

Library of Congress Catalog Card Number 75-4497
International Standard Book Number 0-87168-224-9
AAAS Miscellaneous Publication Number 75-4

Table of Contents

Preface

As science and technology increasingly intersect areas of social and ethical concern, gray areas of professional and personal responsibility emerge. The mounting demand for accountability does not exempt scientists and engineers. Because the texture and boundaries of scientific freedom and responsibility resist simple definition, the position of scientists needs to be reassessed from time to time in the context of changing ethical predicaments.

The Committee on Scientific Freedom and Responsibility was established by the American Association for the Advancement of Science in December 1970 in response to its concern for independent scientific inquiry and responsible scientific conduct. The Committee was directed to study and report on the conditions required for scientific freedom and responsibility, to prescribe criteria and procedures for the study of these conditions, and to recommend mechanisms to enable the Association to review instances of alleged abridgment of scientific freedom and alleged violations of responsible scientific conduct.

The report of the Committee is presented here in the hope that it will illuminate the problems that arise in conflicts involving scientific freedom and responsible scientific conduct and will spur efforts by individuals, professional societies, industrial organizations, and government agencies toward the development of mechanisms for the just resolution of such disputes.

William D. Carey
Executive Officer, AAAS

Acknowledgments

The members of the AAAS Committee on Scientific Freedom and Responsibility are: Allen V. Astin, Director Emeritus of the National Bureau of Standards and Home Secretary of the National Academy of Sciences, Chairman; John T. Edsall, Professor of Biochemistry Emeritus, Harvard University; Walter J. Hickel, former Governor of Alaska and former U.S. Secretary of the Interior; John H. Knowles, President of the Rockefeller Foundation; Earl Warren, former Chief Justice of the U.S. Supreme Court (who died in July 1974); and Dorothy Zinberg, Lecturer in Sociology, Harvard University. Mary Catherine Bateson, associate professor in the Department of Sociology and Anthropology, Northeastern University, served on the Committee until early 1972.

The Committee members wish to record their sense of loss at the death of Earl Warren before this report was completed. Although he did not live to see the final version, we are much indebted to him for wise and helpful advice throughout the greater part of the work of the Committee.

The members of the Committee wish to express their thanks to William Bevan and Richard Trumbull, who were, respectively, Executive Officer and Deputy Executive Officer of the AAAS during the time when most of the Committee deliberations took place. An earlier draft report, although not adopted by the Committee, was prepared by Richard Trumbull, and his efforts are gratefully acknowledged.

Summary

Science and its applications now affect the quality of life for everyone, largely through technological developments that can bring great benefits but can also expose us, and our environment, to grave and often unforeseen hazards. Scientific experts, including engineers, physicians, and many others, are particularly qualified to point out opportunities for technological improvement and to warn of dangers. Experts who may feel an obligation to speak out on such issues may come into sharp conflict with their employers, or with other vested interests. How are such conflicts to be resolved, in the service of truth and the public interest, so that the scientist has both the freedom to speak and the responsibility to inform and warn the public? How are such conflicts to be adjudicated fairly, with safeguards for the rights of all concerned? Science is now inextricably intertwined with major political, social, and economic problems. These impose powerful stresses on many scientists who must deal with political issues that cannot be wisely resolved without the application of expert scientific knowledge. To maintain honesty and objectivity under such pressures is often very difficult, but it is essential if the scientist is to discharge his social obligations and maintain the integrity of science.

Because of its concern over the resolution of such conflicts, the American Association for the Advancement of Science (AAAS) has given the following charge to this Committee: (i) to study and report on the general conditions required for scientific freedom and responsibility; (ii) to develop suitable criteria and procedures for the objective and impartial study of these problems; and (iii) to recommend mechanisms to enable the Association to review specific instances in which scientific freedom is alleged to have been abridged or otherwise endangered, or responsible scientific conduct is alleged to have been violated.

This report deals primarily with the first two of these charges to the Committee, although we also offer proposals concerning the third.

We define the scientific community broadly, to include researchers in both basic and applied science, secondary school and college teachers of science, engineers, physicians, public health workers, technicians, and others who must use some expert knowledge in their work. We consider that the issues of scientific freedom and responsibility are basically inseparable. The responsibilities are primary; scientists can lay claim to special rights, beyond those of other citizens, only when these involve the freedom to discharge their special responsibilities.

The basic function of the scientific community is the advancement of knowledge, including its clarification, interpretation, diffusion, and evaluation. Critical reviewing of material submitted for publication, involving much rejection and revision, is essential to weed out shoddy, trivial, or erroneous material so far as possible; yet the system must endeavor to give a hearing to the daring innovator, whose work might be dismissed by conventional minds as nonsense. Ambitious scientists live often in a state of tension, for they are eager to demonstrate the priority of their achievements yet are urged by the code of scientific ethics to serve the advancement of knowledge in an impersonal and objective spirit. The tension is often inescapable, but the need to maintain uncorrupted standards in seeking and reporting new knowledge is central to the integrity of the scientific enterprise. The temptation to steal the scientific ideas of others, before they are published, is often powerful and may occur without conscious awareness of what is happening; yet it is imperative that the responsible scientist must resist it.

This Committee accepts the advancement of knowledge as a fundamental value; yet certain means for the achievement of knowledge must be either renounced or employed only with special safeguards. This applies particularly to experiments on human beings, that may subject the participants to physical or psychological hazards. These call not only for informed consent on the part of the subjects, but commonly also for careful review by an independent body of experts, sensitive to ethical concerns, who must judge whether the possible benefits from the research outweigh the risks. The American Psychological Association has formulated criteria for ethical standards in psychological research. Our report briefly describes these thoughtful proposals. They provide no simple answers in particular doubtful cases; ethical judgments in this area are still evolving, as concerned individuals learn more and ponder problems more deeply. In rare instances hazardous experiments, like those that established the transmission of yellow fever by mosquitoes, are clearly worth doing, even at the risk of death to the participants. Moreover, we believe that some current restrictions on research have definitely gone too far, for example, the current ban on research on the human fetus, which threatens to prevent the acquisition of knowledge that can greatly benefit the development of normal fetuses and young children. We also maintain the inherent legitimacy of research into the role of genetic and environmental factors in human intelligence and behavior, but we emphasize the responsibility of those involved to seek out, and so far as possible eliminate, sources of personal and emotional bias in planning and reporting their studies.

A group of distinguished molecular biologists has recently called for a moratorium on certain kinds of experiments involving the transplanting of foreign genes into bacteria, lest they give rise to infections in humans that might be dangerous and difficult to control. We applaud this step, at least as a temporary measure, as an example of scientific responsibility.

The open character of science is, in general, essential for the advancement of knowledge and the detection and elimination of error; yet in studies on human behavior it is often imperative to guard the confidentiality of the subjects of the study. Often these subjects will not cooperate, or will not give true answers to questions, unless confidentiality is assured. We consider the difficult problem of obtaining reliable, and independently verifiable, knowledge from studies carried on under such conditions.

Conflicts arising in connection with technology assessment and control are central to the concerns of this Committee. We regard the wise development of innovative technology as one essential factor in solving the formidable problems that the world now encounters; but we must face, and if possible anticipate, the undesirable effects that often accompany, and sometimes dominate, the intended beneficial effects of new systems and devices. Chlorinated hydrocarbon pesticides and the supersonic transport provide two examples of such problems; we discuss others also. Recent sharp disputes concerning standards for protection against ionizing radiation, and their partial resolution by the work of the Advisory Committee on the Biological Effects of Ionizing Radiation (BEIR Committee) of the National Academy of Sciences, furnish one example of an effective approach to such problems. We consider the role of physicians and public health experts in protecting industrial workers and others against poisons and safety hazards to which they are exposed. The prime obligation of such experts is to protect the health of these workers; but, if the health expert comes into conflict with the management in connection with the priority to be given to concerns over health, his position is weak unless he has independent status or the strong backing of an independent professional organization.

The criteria for resolving such conflicts involve human health and well-being, environmental protection, and the evaluation of the benefits derived from industrial production in relation to the human and environmental costs involved. Considerations of justice are important; it is, for instance, unjust to expose industrial workers to the risks of grave disease and premature death in order to make a cheaper product for consumers and enhance the profits of the manufacturer.

We emphasize the importance of due process in dealing with disputes arising from "whistle blowing" by an expert who warns of the dangers of some substance or process. Some 20 engineering and scientific societies have recently promulgated Employment Guidelines, which represent an important step in this direction; but, in addition to these guidelines, a clear procedure for the arbitration of disputes is necessary. This would involve hearings before a board with independent members, and also some right of appeal. We believe that the professional societies should assume a greatly increased responsibility for their members, in conflicts that involve defense of the public interest on issues related to the professional competence of the members of the society. The difficulties associated with this proposal, and possible ways of solving them, are discussed. Some of these problems are illustrated by the case of the three Bay Area Rapid Transit (BART) engineers in San Francisco who warned, correctly, of dangerous deficiencies in the BART automatic control system and later were fired as "troublemakers."

The role of the AAAS, which is a comprehensive organization with many constituent societies, would not permit it to handle the numerous cases of conflict that are more properly in the area of responsibility of the individual societies. The Association, however, can play an important coordinating role between the societies in helping to formulate codes of ethics. It is already publicizing many alleged cases of abridgment of scientific freedom in *Science*; it can extend this role, and also serve in important cases as a forum for the expression of conflicting viewpoints. The annual meetings of the Association also provide such opportunities. In certain cases the Association may set up a special commission to deal with a problem of unusual importance, as it did with the study of the effects of defoliants and herbicides in Southeast Asia. Finally, we emphasize that problems of scientific freedom and responsibility are continuing problems; they may call for a reevaluation of the situation by this or another committee some years hence, or for the establishment of a "watchdog" committee to receive complaints and refer them, if possible, to some other appropriate body when they appear to have merit.

Chapter One

Introduction

Scientific Freedom and Responsibility

Problems of scientific freedom and responsibility are not new. They arose in acute form, well over three centuries ago when Galileo was brought to trial for his heliocentric views, in an intellectual and spiritual climate very different from our own. They have arisen many times since then, notably in the controversy over Darwin and evolution a century ago. During the last 30 years such problems have become incomparably more numerous, more urgent, and more complex than ever before. Faith that scientific and technological progress was supremely desirable has given way to far more troubled and more ambivalent attitudes. Scientific research has penetrated deeply into the nature of atoms, of radiation, and of the subatomic world; it has unraveled the mysteries of the genetic code; it has begun to illuminate the workings of the brain and the nervous system. It is thereby altering, deepening, and enriching our outlook on the nature of the universe and of life. It has helped to liberate mankind from many fears, notably, in large measure from the terrors of infectious disease. The need for further development of science and technology is intensely felt today; government support for research and development, in spite of cutbacks in the last few years, is greater by several orders of magnitude than it was in 1940.

Yet some of the consequences of science and technology frighten, discourage, and even horrify most of us. Nuclear weapons, which developed rapidly as the result of an unexpected discovery in basic science, have the capacity to wreck whole civilizations in a matter of hours. Less well known, but possibly of comparable gravity, are the dangers from some of the creations of the organic chemists and biochemists. For instance, some of the chlorinated dioxin derivatives which appear as unwanted minor byproducts in the preparation of the herbicide 2,4,5-T have been found to be lethal to guinea pigs at concentrations of 600 parts per trillion (*1*). Their toxicity for human beings is not known, but compounds of this class appear to attack some very fundamental aspect of cellular activity. The development of other, still more toxic compounds remains a live possibility. If deliberately manufactured and used for destructive purposes, they could be as dangerous as nuclear weapons.

The unprecedented rate of population growth, largely a consequence of advances in medical science, sanitary technology, and transportation, threatens the ecological balance of the earth and raises the specter of catastrophic famines. These dangers are so well known that only a reminder of them is needed here.

These and other urgent problems, in which science and its applications are profoundly involved, have led to intense and often bitter controversy among members of the scientific community, and between scientists and policy-makers. We need only recall, as examples, the controversies over the biological hazards of nuclear weapons testing, over the use of herbicides and defoliants in the Vietnamese war, over the effects of DDT and other chlorinated hydrocarbon pesticides, over the development of an antiballistic missile program, over the supersonic transport, and over the extent of the hazards from nuclear power plants.

In these controversies there are commonly powerful forces, in government, in industry, or in both, that support one side of the issue under dispute. Opponents of these programs have often believed that they lacked an adequate opportunity to be heard, or lacked the funds and facilities to gather evidence regarding the issues under dispute. In some cases, moreover, scientists who wished to raise these issues were employed by an organization which, they believed, was obstructing consideration of what they held to be weighty scientific evidence. Sometimes the protesting scientists became convinced that the organization was trying to muzzle them and suppress their findings, and perhaps would even take reprisals against them. In other cases powerful and well-funded organizations have *opposed* programs of substantial technical and social merit, while the unorganized technical experts felt similarly the lack of access to data and facilities.

The American Association for the Advancement of Science (AAAS) adheres strongly to the position that such controversial issues should be debated with the fullest possible presentation of the scientific evidence, and that the resulting decisions should be made with due consideration of all the evidence. Yet it would be far beyond the power of the AAAS to act as a court of appeal in such disputes, except in rare and quite extraordinary cases. However, because of its deep concern over these issues, the Board of Directors of the AAAS has charged this Committee to consider the following matters:

1) To study and report on the general conditions required for scientific freedom and responsibility.

2) To develop suitable criteria and procedures for the objective and impartial study of these problems.

3) To recommend mechanisms to enable the Association to review specific instances in which scientific freedom is alleged to have been abridged or otherwise endangered, or responsible scientific conduct is alleged to have been violated.

In this report we deal primarily with the first two of these items in the charge to the Committee. We also offer some considerations relating to the third item but have concluded that a more complete answer would be beyond our competence. We note, with regard to the second item, that some of the various professional societies of scientists, engineers, physicians, and other related groups are now beginning to assume a much more active role than in the past, in defense of the rights and responsibilities of their members to speak out on controversial issues that arise from their professional work and their expert knowledge. In a later section of this report we discuss in more detail the opportunities of the professional societies to serve in promoting the balanced and objective consideration of these controversial issues and also the limitations to which such societies are subject.

The role of the AAAS, as a comprehensive organization covering all branches of science, with numerous professional societies as affiliates, must necessarily be different from that of a more specialized professional society. Clearly, however, it has an important role to play in publicizing important controversial issues, in giving a hearing to the conflicting views of the participants in such controversies, and on very rare and special occasions in appointing a committee of inquiry to hear the conflicting evidence and report its conclusions publicly. We give further consideration to these matters in the final section of this report.

Chapter Two

The First Charge

The Scientific Community: Its Rights and Responsibilities

The American scientific community, as we define it here, includes a wide range of very diverse individuals, who number at the very least several hundred thousand and possibly more than a million. It includes basic scientists, in universities and research institutes, and applied scientists, engineers, physicians, and others working in the medical sciences for whom a knowledge of science and how to apply it is an essential part of their activities. It includes graduate students and technicians working in scientific laboratories and on scientific problems and also teachers of science, whether involved in research or not, both in colleges and in secondary schools. In what follows we shall speak primarily of the rights and responsibilities of scientists involved in research, both basic and applied; but much of what we say is applicable to the scientific community as a whole. On occasion we distinguish the roles of basic scientists from those of other workers whom we denote as technologists. We do this simply to distinguish their primary functions; we consider these two categories as having equal rank, although different functions, as members of the scientific community.

Although the later sections of this report do not deal with the rights and responsibilities of school teachers, we are not unmindful of their problems. For instance, certain groups in California have recently attempted to require biology teachers to teach the doctrine of special creation of species side by side with the concept of evolution, giving equal weight to both (*2*). Clearly such a requirement would violate both the freedom and the responsibility of the teacher to present the scientific evidence to his or her students as honestly as possible, in the light of all the available facts.

The Committee concluded, early in its deliberations, that the issues of scientific freedom and responsibility are basically inseparable. Scientific freedom, like academic freedom, is an acquired right, generally approved by society as necessary for the advancement of knowledge from which society may benefit (*3*). The responsibilities are primary; scientists can claim no special rights, other than those possessed by every citizen, except those necessary to fulfill the responsibilities that arise from the possession of special knowledge and of the insight arising from that knowledge.

Later we shall have much to say of the activities often referred to as "whistle blowing" in which issues of freedom and responsibility are inextricably intermingled. Whistle blowing involves situations in which a scientist, engineer, physician, or other expert, becomes aware of hazards arising from some process, material, or product, or, on the other hand, becomes aware of possible improvements in technology or procedure that deserve to be adopted but are being neglected. Issues of public safety are frequently involved, and often the whistle blower works for the marketer of the process or product. Some may argue that persons with expert knowledge have a "right" to release information in their possession, if such release is in the public interest. Others would say that it is the responsibility of such experts to release the information, even though they might prefer to remain silent. Both rights and responsibilities are clearly involved here, but it seems clear to us that the responsibilities are primary.

Presumably the potential whistle blower will begin by reporting his concern to his employer and urging that corrective measures be taken; if the matter can be settled without appeal to outside authority, so much the better. If this procedure fails, however, and the concerned employee decides that he has the responsibility to make the matter public, he faces obvious risks that may include the loss of his job. If whistle blowers are to be encouraged to take such risks—and we believe that they should be encouraged when serious issues are involved—they must be assured of some form of due process in passing judgment on the issues that they raise. This would call for the presence of independent outside members on any board that passes judgment on the issues, and should also include some right of appeal. We shall return to these matters later.

Science and Values

It is often said that science is ethically neutral and value-free. Such statements, in our opinion, are seriously misleading and in some respects quite false. It is, of course, obvious that a scientific discovery, once published, is available for anyone to use; it can be used in exceedingly diverse ways, with consequences that may be good or bad, or commonly a complicated mixture of both. The *activities* of scientists and technologists, however, are conditioned and directed at every turn by considerations of human values. This is true over the whole range of activity, from the most basic research to the applications of science in technology. The patterns of value that are involved do in fact differ widely at the two ends of the spectrum, although there is no inherent conflict between them.

Patterns of Value and Responsibility in the Basic Sciences: The Maintenance of Scientific Quality and Integrity

The men and women who are involved in the advancement of basic scientific knowledge are to be found, for the most part, in universities and in certain research institutes; many are in industrial and governmental laboratories, and a few are scattered elsewhere or work independently. In all these places one may find workers in applied science and technology as well; indeed, the same person may be at one time concerned with basic science and at another with its applications to a technological problem, and he may pass from one to the other with no sense of discontinuity.

There are, however, certain patterns of values and behavior particularly characteristic of the community of basic scientists. We attempt to describe them briefly here. In a number of recent studies investigators of the activities of this portion of the scientific community have illuminated the subject (*4–11*), and we note here our indebtedness to them in what follows.

The central value for the most creative scientists is a passionate dedication to the advancement of knowledge, and to the solution of the baffling puzzles with which nature confronts us. The desire, for instance, to discover the principles underlying quantum phenomena at the atomic and subatomic level, or to understand the chemical nature of the gene and its replication, has preoccupied some of the greatest scientific minds of this century. Such workers often live with their problems, day and night, for months or even years on end; the desire to solve a particularly baffling problem may keep them preoccupied, to an extent that most other people can scarcely conceive. A deep sense of the beauty of science, and its aesthetic appeal, commonly accompanies this urge to solve difficult and fundamental problems. Indeed, the aesthetic appeal of science is very strong for many other, less creative researchers; but this appreciation of the beauty of science requires in general an arduous preliminary training that makes it unattainable for most nonscientists.

We do not wish to idealize the character of scientists. Many fields of research today are feverishly competitive, and some brilliant scientists are ruthlessly determined to get ahead of their competitors, sometimes by unscrupulous techniques. The extent to which this is true varies greatly with the character of the individual researcher, and with the current state of the field of research. We must note that some of the greatest scientists, such as Newton, have been extremely jealous and suspicious of their rivals.

Most working scientists are less deeply preoccupied with their work; for many, the chief satisfaction in work comes from doing a skilled job and solving some new but limited problems on the way. This is true, for instance, of many organic chemists who synthesize new compounds of a class whose general characteristics are already well known, or of biochemists who refine the details of some metabolic pathway that is already known in outline. Such work is absolutely indispensable for the coherent development of the great body of science; and from time to time, of course, it reveals new and unexpected phenomena that form the starting point for entirely new lines of research.

In any case, scientific researchers, if they are to contribute to the advancement of knowledge, must operate with a high degree of flexibility and freedom to follow unexpected leads. Administrators of organizations devoted to basic research must therefore give the researchers wide latitude in pursuing their inquiries. Attempts to fit such research into organized programs are often self-defeating. This does not mean that basic research proceeds in isolation, unaffected by social values and needs, or that it is always damaged by attempts to manage or organize it. On the contrary, particularly since most such research today involves support from government funds, the directions taken by fundamental research are profoundly affected by the general values prevalent among the sponsors of the research. The working scientist is often unconscious of these factors and commonly pursues his work with an inherent concern for the advancement of knowledge as he sees it, and occasionally with new insights that may lead to unexpected and surprising vistas. The investigator who is following a trail of research into unknown territory will be guided in his approach largely by the inherent intellectual challenge of the subject; but the initial decision to start on one trail rather than another is likely to be greatly affected by current social values that tend to establish what is worth doing.

In the activities of scientists there is a constant interplay between imaginative guessing and the formation of hypotheses, on the one hand, and rigorous critical testing of hypotheses, on the other. Neither activity alone is sufficient; it is the combination of the two that gives science its capacity for continual advance, for the rejection of imperfect hypotheses and their replacement by better ones, which in turn are subject to critical examination and possible refutation. These processes have been most searchingly analyzed by Popper (*12*); Medawar (*13*) has presented them more simply and with great lucidity.

One of the basic responsibilities of scientists is to maintain the quality and integrity of the work of the scientific community. Ideally, it is an open community—all findings should be publicly and generally available, and open to criticism, improvement, and, if necessary, rejection. (We discuss later the special requirements for confidentiality that arise in connection with some kinds of research on human beings.) Publication of research, indeed, is generally conditioned on a rather elaborate process of critical examination and review. Authors submit papers to editors, editors send them to referees, and the referees commonly subject the papers to searching criticism. The referee system has in practice proved indispensable in the endeavor to minimize the publication of trivial or erroneous work and to maintain standards of clarity and compactness in the flood of papers published (*14*). The system works imperfectly; much that is erroneous, and more that is trivial, appears in print. The danger also exists that uninformed or prejudiced reviewers may reject work of great originality and importance. A classic example is that of J. J. Waterston, whose profoundly original paper on the kinetic theory of gases, submitted in 1845, was rejected by the Royal Society referees as "nothing but nonsense." Waterston, disheartened, abandoned science altogether. [See Merton (*4*), pp. 318 and 456; and Ihde (*15*), p. 401.] In 1892 Lord Rayleigh found Waterston's manuscript in the files of the Royal Society and had it published; by then, of course, Clerk Maxwell and others, knowing nothing of Waterston's work, had independently developed the theory much further. It is conceivable that some referees today are making similar shocking blunders. Usually, however, such new and unconventional work, under present conditions, finds an outlet in one place or another; but we should not assume complacently that this is always the case.

Communication and Theft of Scientific Ideas

The lifeblood of science is communication. Scientists frequently exchange ideas and discuss with their colleagues the problems of their current work. Yet they also have a strong sense of personal property in their original ideas and discoveries, which form the basis for their position and prestige in the scientific world. Inevitably there are opportunities for the conscious or unconscious theft of ideas; this can, of course, give rise to intensely bitter feelings (*16*).

There can be, on occasion, great difficulties in the relations of senior and junior investigators in the same laboratory or department. A junior investigator in a research group may make a major discovery, for which his research supervisor may then assume credit. On more than one occasion the AAAS has received an appeal from a scientist who considers himself to be the victim of such injustice. He may present extensive documentation to back up his claim. These are grave matters; but it is generally extremely difficult to see that justice is done in such a case. Often no documents exist that could establish what really happened; the crucial events may have involved informal conversations of which no record exists. On rare occasions, when the discovery involved is a truly major one and the evidence of injustice is strong, the AAAS might well decide to publicize such a controversy, permitting both parties in the dispute to state their cases briefly, with suitable references to documentary evidence for those who wish to inquire further into the dispute.

Referees of scientific papers are exposed to the temptation to steal ideas from the manuscripts they review. A referee may be placed in a very embarrassing position when the paper to be reviewed is closely related to his own research. He may be about to submit a paper for publication himself, on an almost identical subject. If he learns, as a referee, that a rival has anticipated him, and if he judges the work to be valid, it is his obvious obligation to return the paper promptly to the editor and recommend its publication. The referee may, however, have valid criticisms of his rival's work, which would call for revision by the author and perhaps for more experiments. What should he do about reporting his own work—rush it in quickly for publication, proceed as if nothing had happened, or quietly bury his own work on the ground that reporting it would now be superfluous? Some highly scrupulous, perhaps overscrupulous, scientists will follow the latter course; some who are unscrupulous will seek to delay publication of a rival's work and get their own work out first. Obviously such action is to be strongly condemned, but even conscientious scientists can be faced with painful dilemmas in such situations.

Similar problems face members of panels who review applications for research grants. It is very easy, while reading an application for a grant, to pick up an idea from it and subsequently to think of it as one's own. Here again it is impossible to lay down precise rules of conduct; the theft of another person's ideas may be subtle and unconscious. Members of such panels are aware of the problem and must endeavor to be scrupulous. The basic fact is that the system of peer review of research grant applications has worked exceedingly well. There is no adequate substitute for it, in maintaining support for work of high quality; and abuses have apparently been rare.

It is vital to maintain the continuity of the scientific enterprise: hence great stress is laid on the requirement that scientific investigators give due credit to their predecessors by proper citation of their work. We have already spoken of the importance of contributions made by junior researchers being recognized by the senior investigator with whom they are working. Recognition for original contributions is the chief reward and honor for the scientist in basic research; hence there are often passionate feelings about priority, which concern many of the greatest scientists as well as minor contributors. This has been documented extensively by Merton [*(4)*, chapters 14 and 15]. Among many scientists there is an inner conflict between the desire for recognition and priority and the sense of obligation that scientific endeavor be viewed as the search for impersonal truth. The latter ideal serves to temper and restrain the self-assertiveness of the individual scientist, but for many gifted workers the inner conflict remains.

Freedom of communication among scientists is essential for scientific progress, and for the critical validation or invalidation of scientific findings. This includes freedom to attend meetings, to visit colleagues, and to write and telephone them. Americans, for the most part, take these freedoms for granted; but in many countries drastic restrictions exist, as Medvedev (*17*) has documented in detail for the Soviet Union from his own experiences. Even in the United States there have been disturbing episodes; in the early 1950's, for instance, the U.S. Public Health Service revoked a number of unclassified research grants and declined to support others because of unevaluated adverse reports in the security files of these investigators (*18*), of which the investigators themselves remained ignorant. This practice was ended after a review of the situation by a committee of the National Academy of Sciences (*19*). More recently it was discovered that the Public Health Service had a secret blacklist of scientists who were disapproved for service on review panels for research grants, even though the research involved was completely open and unclassified, and some of the investigators were highly eminent. Again, this practice was ended when the facts became public (*20*).

Hostility to Science

In this discussion of basic science we have assumed, as a fundamental postulate, that the advancement of scientific knowledge of the world, including man, is inherently good and desirable. There have been civilizations in the past, and there are perhaps some today, that are hostile to this spirit of free exploration and inquiry. Some would subscribe to the views of the Preacher in Ecclesiastes (1:18): "He that increaseth knowledge increaseth sorrow." Even in technically advanced countries, such as the United States today, there is an undertone of fear of science, and hostility to it, among significant sections of the community. Many, perhaps most, citizens welcome science only because they expect it to bring practical benefits to their way of life, and have little or no appreciation of the driving motives of the research scientists themselves. Many citizens have deep and legitimate apprehensions concerning many aspects of applied science and technology; of these we shall speak in detail later. There are also sensitive, and sometimes artistically gifted, people who regard the rigorous and austere character of scientific knowledge as dangerously inhuman, or even antihuman, and contrast it with the direct and instinctive knowledge derived from spontaneous communion with man and nature, without attempts at abstract intellectual analysis (*21*). Few scientists, we think, would feel that science is so inhuman, but they do frequently have great difficulty in conveying to others the sense of the beauty and grandeur of the perspective on the world that science imparts. It is part of the responsibility of scientists to convey, as far as possible, the nature of that vision to the world at large.

Should There Be Forbidden Areas in the Realm of Basic Research?

Those for whom the advancement of knowledge is a supreme value might well believe that nothing should be off limits as a possible subject for basic research, as distinct from applied science and technology. Yet there may be strong reasons for declaring certain areas of research off limits, or at least for imposing severe restrictions on them. Research involving human subjects inevitably raises serious questions: we generally acknowledge today the need for obtaining informed consent from the subjects of such research. In some situations genuinely informed consent may be difficult or impossible to obtain—prisoners, for instance, are frequently used in the testing of new drugs, because they are confined and because they are commonly willing to serve as subjects for much less pay than people outside of prison would generally demand. They are inevitably subject to pressures that most of us do not feel. Experiments on very young children, who are in no condition to give informed consent, are done with the consent of their parents; this obviously involves a heavy responsibility, both for the investigator and for the parents.

There are certain experiments that can cause serious physical or mental suffering to the participants, and in extreme cases may even endanger their lives. Others, particularly a few of the psychological experiments that have been done or contemplated, might be regarded as morally degrading or psychologically damaging to those involved. Experiments which cause suffering in animals are also subject to serious question, and should be undertaken only for extraordinary reasons. Some studies involving deliberate ecological modification of regions of considerable size are now being considered. In such cases, one must weigh the prospective value of the information that may be gained from the experiment against the harmful effects of possible damage to the region involved.

Some experiments are justified, even if they involve great risks to the participants. The experiments that conclusively demonstrated the transmission of yellow fever by mosquitoes, about 75 years ago, furnish an outstanding example. The doctors who allowed themselves to be bitten by infected mosquitoes knew that they were risking their lives. One of them, Dr. Jesse Lazear, did die of yellow fever, and others became gravely ill. The control group, who slept in the beds just vacated by yellow fever patients and were exposed to every possible source of infection except the mosquitoes, also took what might have been great risks though in fact they did not suffer. Such heroic experiments are fortunately seldom called for, but they may well be needed from time to time in the future. In such cases, it is clear that the participants must completely understand and accept, in advance, the risks that they are to take.

Recently, in a statement probably unprecedented in the history of science, an eminent group of researchers, headed by Paul Berg of Stanford University, has deliberately renounced, for the time being, the performing of certain experiments of probably great scientific interest, because of potential though unproven hazards to human health (*22*). The proposed experiments would involve the use of some newly discovered enzymes, which serve to introduce genetic material of other species into bacteria and other living cells. The experiments that these researchers are renouncing are of two kinds: (i) the introduction into bacteria of genetic material from other strains or species of bacteria that might confer resistance to antibiotics or produce bacterial toxins; and (ii) the insertion, into bacterial DNA, of foreign DNA from cancer-producing viruses or other viruses, which might then become widely disseminated among bacterial populations in humans and other species. A third class of experiments, which would involve inserting animal genes into bacteria, is not specifically renounced but is considered risky: "such experiments should not be undertaken lightly," says the statement (*22*).

We note that this distinguished group of scientists speaks as the Committee on Recombinant DNA Molecules of the Assembly of Life Sciences of the National Academy of Sciences. They urge the Director of the National Institutes of Health to establish an advisory committee to supervise possible research in this area, and to control and minimize potential hazards. They also urge the holding of an international meeting of involved scientists from all over the world, to review scientific progress in this area and recommend ways of dealing with potential hazards.

Clearly this declaration represents a landmark in the assumption of scientific responsibility by scientists themselves for the possible dangerous consequences of their work. The scientists involved emphasize that the hazards they foresee are as yet only potential, not demonstrated; as further knowledge is gained, workers in this field may have reason to recommend relaxing the ban that they have temporarily imposed upon themselves. Obviously, there are at present no mechanisms to compel other scientists, who may wish to work on these problems, to refrain from experimenting in this area. However, the force of moral disapproval by a group of eminent scientists may act as a powerful deterrent to experimentation by others. At the time of the writing of this report, it is too early to tell *(22a)*.

The decision in this instance to refrain temporarily from further experiments is based on possible direct hazards to human health. For several years past, some individual scientists have warned on other grounds of possible dangers from the development of "genetic engineering," including such processes as the production of multiple copies of genetically identical individuals by cloning. The suggested threats, in these matters, are not so much to health as to human integrity, dignity, and individuality. It seems to us that we should be on the alert for such possible threats, but we see no justification, at the present time, for any attempt to impose restrictions on the freedom of such research in genetics. The dangers, if they exist, are remote and in our opinion are decisively outweighed by the great benefits that such research can bring to mankind.

The Legitimacy of Research on the Role of Genetic and Environmental Factors in Human Behavior

Another area of research that some have suggested should be off limits for scientific inquiry involves the role of genetic and environmental factors in determining the attainable level of human achievement in different individuals and groups. The debate has tended to center on the use and interpretation of IQ tests, although the total problem of human capacity and performance—and indeed the concept of intelligence—is obviously far wider than this. The bitterness that has attended much of the discussion in this field obviously arises from the fact that blacks in general score distinctly lower in IQ tests than whites from comparable social and economic levels. Some investigators, notably Arthur Jensen (*23*), have drawn the conclusion that the differences between racial groups in average performance on these tests are due primarily to genetic factórs. Others have naturally hotly disputed such claims, and indeed some eminent scientists have argued that research done in this field serves to promote "racism," and that therefore it should be condemned by the scientific community.

We reject the latter proposal. It would appear to us to be a goal of supreme scientific and practical importance to decipher the full range of genetic potentialities and their variations between different human beings, and the modes by which these potentialities either reach their full development or are inhibited and thwarted by unfavorable environmental factors. The evidence that human intelligence, and capacity for achievement, involves a large genetic component appears strong [for a good brief summary of the evidence, see, for instance, Dobzhansky (*24*), but see also the recent report by Jensen (*23*) of the many strange anomalies in the work of Sir Cyril Burt, on which much of the presumed evidence for the heritability of IQ has been based]. Conversely, there is also powerful evidence that serious malnutrition in early childhood, or malnutrition of the mother before the birth of the child, can do irreversible damage to the native intellectual capacity of the child and prevent the realization of the full genetic potential of the individual. There is also evidence, extensive although less clear-cut, that adverse social and psychological factors in early childhood can exert a similarly crippling influence. The problems of research in this area are appallingly difficult, and they raise grave ethical issues. It would clearly be immoral to subject young children deliberately, on an experimental basis, to malnutrition or to various forms of social suffering and deprivation in order to observe their behavior in comparison with that of a carefully chosen group of controls. Some related studies on animals, notably the work of H. F. Harlow and his associates on the effects of psychological deprivation in young monkeys, have yielded valuable insights, with important implications for human development. With animals also, however, one must consider carefully the suffering that might be inflicted on them when weighing the question of whether the experiment is justified or not.

It is obviously a formidable task to disentangle the relative roles of genetic and environmental factors in human achievement, but it is profoundly important that we enlarge our understanding in this field, if we are to deal wisely with individual human beings in all their diversity. In view of the immense difficulties involved in obtaining really significant results, it is wise for investigators to state their results cautiously and modestly, to specify clearly the limits of accuracy, and to refrain from sweeping claims based on highly specialized researches. Disentangling the evidence concerning variation among individuals is difficult enough; the attempt to draw conclusions concerning the relative roles of heredity and environment among different racial groups is a far more uncertain business. It is certainly premature today to draw conclusions concerning social policy from the present state of research in this field. A recent statement by Harvey Brooks, concerning a very different problem, is highly relevant here. "Scientists can no longer afford to be naive about the political effects of publicly stated scientific opinions. If the effect of their scientific views is politically potent, they have an obligation to declare their political or value assumptions, and to try to be honest with themselves, their colleagues, and their audience about the degree to which these assumptions have affected their selection and interpretation of scientific evidence. Once scientific opinion enters into the public domain, the possibility of political neutrality disappears, but this does not mean that objectivity should be thrown to the winds rather than being held up as an ideal to be achieved as closely as circumstances allow." (*25*).

We reaffirm, however, the view that inquiry into genetic differences between racial groups is a thoroughly legitimate field of research, and we condemn the emotional attacks and personal threats that have been made against such persons as Arthur Jensen and R. J. Herrnstein, because of their views concerning the importance of genetic factors in human intelligence. Their views are, of course, open to critical debate and challenge on scientific grounds, as long as the debate is conducted on an objective scientific level. Here, as elsewhere, unceasing critical examination of hypotheses is the lifeblood of science.

Problems of Informed Consent and Ethical Principles in Psychological Research

Many of the problems that set limits to the allowable domain of research appear with particular force in the field of psychology. Psychological experiments, like many medical experiments on human subjects, not infrequently involve some risk of harm to the participants. Also such experiments frequently, by their very nature, involve some deception of the participants as an integral aspect of the study.

Temporary deception is often inherent in the design of the experiment, which would be pointless if the subject knew all about it in advance. Yet the use of deception inevitably raises ethical issues. Recognizing the seriousness of these problems, a committee of the American Psychological Association has recently prepared a detailed report on ethical principles in the conduct of such experiments (*26*). They state ten general principles, which involve the personal responsibility of the investigator to evaluate the ethical acceptability of a research project; his responsibility for his collaborators, subjects, and students, who are, however, not thereby relieved of responsibility themselves; the need to inform participants in advance of all aspects of the research that may influence their willingness to participate; the problems raised by the use of deception in experiments; the right of individuals to decline to participate in the experiment and to withdraw from participation at any time; the protection of participants from physical and mental discomfort, harm, and danger; the need to inform participants, after the experiment, of the nature of the study, in order to remove misconceptions; and the responsibility of the investigator to detect and remove, or correct, any damaging consequences of the experiment. They emphasize also the importance of maintaining the confidentiality of the data obtained in the experiment, insofar as it pertains to particular individuals. They illustrate all these principles in terms of specific examples, where researchers have had to make difficult decisions. They point out that frequently there is a conflict between the value of the knowledge that may be gained from scientific experiments and the risk of damage to the participants and the possible violation of ethical principles. In general, they carefully avoid dogmatic statements on what is or is not allowable, recognizing that the individual investigator must bear a heavy responsibility for difficult decisions. In doubtful cases, however, the eager investigator is naturally likely to be biased in favor of going ahead with an experiment that he has planned, even when its justification might seem dubious to others; in doubtful cases, therefore, it is essential for him to seek advice from others, perhaps from duly constituted panels set up to deal with these issues.

Psychological experiments represent, perhaps in an extreme form, the class of ethical problems that arise in experiments on human subjects, and the statement by this Committee is a particularly thoughtful attempt to formulate the guiding principles. It is by its nature tentative. Modern science is venturing into much unknown territory here; we are faced with a conflict of values between the advancement of knowledge and the protection of the individual subject, and the ethical principles involved are still in the process of formation.

When we pass from this careful statement of principles to the realm of applied psychology, we encounter some horrifying evidence of the use of modern psychological research results as tools of torture. In Portugal, for example, under the Caetano regime, recruits were systematically trained in the use of torture (*27*) by the secret police (DGS). According to a report in the *Manchester Guardian*, films and other records showed that doctors took part before, during, and after torture sessions, and kept photographic records for scientific purposes. Included among the doctors were at least two psychiatrists "who supervised the application of the most sophisticated method of sensory deprivation and the denial of sleep" (*27a*). Dr. Alfonso de Albuquerque, a British-trained psychiatrist who has been compiling a dossier of the evidence of torture and has treated some of the patients after their release, said that years later some of them had not recovered from the effects of torture (*27*).

According to a recent report of the London-based organization Amnesty International, summarized in a recent article in the *New York Times* (*28*), torture has now become a state institution in more than 30 countries: "a rule of pain carried out by technicians, scientists, paramilitary officials, judges, and cabinet ministers." What Amnesty International calls a "cancerous growth" of torture has occurred particularly in Latin America, where 22 nations are involved. "Refinements" have resulted from technical and medical research devoted to developing techniques for intensifying pain without causing death or (perhaps) irreversible physical damage. Apparently there exists an international network of schools training military men and technicians in techniques of torture.

Scientists have long been aware of the horrifying possibilities of misapplication of their researches. In these instances the possibilities have evidently become realities. The community of scientists cannot remain indifferent to these developments. Certainly any doctor or psychologist who helps to promote torture should be immediately subject to expulsion from any scientific society to which he belongs, and, if the evidence were clear, he should be publicly blacklisted by his colleagues throughout the world. Such measures would still be grossly inadequate to stop the spread of torture, but the conscience of the scientific community would seem to call for such steps as at least a minimum response.

On the Retardation of Needed Research by Excessive Scruples: The Ban on Fetal Research as an Example

In the preceding section we have discussed the endeavor to balance the need for new knowledge against the risks to participants in experiments that can yield such knowledge. At present, there are powerful pressures in favor of safeguarding experimental subjects, even if this may mean sacrificing the attainment of knowledge that could lead to great human benefits. The complexity of the issues involved emerges from the diverse points of view collected and set forth in a recent comprehensive book (*29*) and in a symposium (*30*). There are no simple answers, as anyone who consults these discussions will realize. In safeguarding possible subjects from harmful procedures, Barnes (*30*) points out the importance, not only of informed consent on the part of the subject, but of careful peer review of the research project before it is approved. In many cases, peer review may be by far the more effective of the two mechanisms for the protection of possible subjects from ill-judged or irresponsible experimentation.

We believe that the imposition of restrictions on research has in some important instances gone too far. An outstanding current example is the ban on fetal research imposed by the National Research Act of 1974 (T.L. 93-348). Fortunately, this ban, under the Act, extends only into part of 1975, but there are strong pressures, in Congress and elsewhere, to impose a permanent ban. Under the current Act it is illegal to experiment on any "living" human fetus, either before or after induced abortion, except in the very unlikely event that the experiment is intended to save the life of that particular fetus. The meaning of "living" is not quite clear, but this would seem to ban any research on a fetus with a beating heart (*31*).

We strongly oppose such restrictions. Research on the human fetus, over the past two decades, has yielded major benefits for human health. Behrman (*32*) has pointed out how many diseases that we can now diagnose, or treat, or both, would have been unmanageable if a ban on fetal research had been in effect. The most notable example is Rh disease, which is now totally preventable, thanks to prolonged research over more than a generation, which often involved the fetus as well as the newborn infant. In addition, there are currently 51 diseases that can be diagnosed *in utero*, thereby allowing appropriate preparation for immediate treatment on delivery; such advances would have been impossible without research on the fetal state. Infections, which take such a large toll during the fetal period and just after birth, require at some stage the testing of new antibiotics on the fetus. Yet in Boston a group of doctors, who were doing such antibiotic studies on fetuses obtained after legal abortions (the mother having received the antibiotic beforehand), have been subjected to legal prosecution by the local district attorney.

The human fetus is extremely susceptible to many drugs, as the thalidomide disaster tragically demonstrated. Much research has shown also that the fetus is far more easily damaged by radiation than the adult. Research on fetuses that do not survive abortions helps us to discover how to give protection from such harmful agents to thousands of other fetuses that are destined to survive and grow up. Given the great contributions that such research can make to the future health of children and adults, we would urge that such research be not only permitted but intensified, subject to careful peer review of the research projects involved.

The Conflict between Science and Secrecy; Problems of Confidentiality in Research on Humans

Many scientists spend much or most of their working time in activity classified as secret. Some work for industrial firms that jealously guard trade secrets; others work for the government, and their findings are classified on grounds of national security. In many cases, the reasons for secrecy are doubtless compelling. Nevertheless, science inevitably suffers from the imposition of secrecy on a research project. The reasons were succinctly stated in 1965, in the report on "The Integrity of Science" (*33*) by the AAAS Committee on Science in the Promotion of Human Welfare [(*33*), p. 177]:

> Free dissemination of information and open discussion is an essential part of the scientific process. Each separate study of nature yields an approximate result and inevitably contains some errors and omissions. Science gets at the truth by a continuous process of self-examination which remedies omissions and corrects errors. This process requires free disclosure of results, general dissemination of findings, interpretations, conclusions, and widespread verification and criticism of results and conclusions.

That report illustrated these general considerations in terms of a discussion of some large-scale experiments in which secrecy led to unfortunate results through lack of awareness of essential facts, facts that might readily have been discovered in advance if free public interchange of the data on the planned experiments had been permitted within the scientific community. We reaffirm the general conclusions of that report: secrecy almost always impedes scientific progress; in applied science and technology it also sometimes permits the development of hazards to the public that could be eliminated if the information were openly available. We hold that, with rare exceptions, data representing a real advance in fundamental science should never be kept secret, except in a major war situation, such as the early stages of the development of the atomic bomb in World War II. Even in such cases scientific information should remain classified only for a limited and specified time; it should then be released automatically, unless a strong and specific case can be made for withholding a particular piece of information for a further limited period of time. We should look with a critical eye at claims of secrecy on grounds of national security. Although some such claims are no doubt valid, it is all too evident that they often serve to cover up evidence of governmental ineptitude or corruption. The events of the last 3 years serve only to emphasize this enduring truth.

When national security is not involved, we can be even more emphatic. A recent example illustrates the point. In 1972, the President's Advisory Panel on Heart Disease responded to an urgent request from the President "to determine why heart disease is so prevalent and so menacing and what can be done about it." Dr. J. H. Comroe, Jr., a member of the Advisory Panel, described what happened. By extraordinary effort the panel submitted a report on 1 September 1972, having first met on 18 May of that year. Within 6 weeks, Dr. Merlin K. Duval, of the Department of Health, Education and Welfare, transmitted it to the President; but it has never been made public. It was classified as confidential and never released, although it was supposed to be urgently needed for planning a national attack on the problems of heart disease. Apparently no member of Congress had seen it, at least by early 1974, when Dr. Comroe's statement on this episode appeared (*34*). The total working time for the experts involved in preparing the report was estimated at 5,000 working days, and it cost several hundred thousand dollars of taxpayers' money. We agree with Dr. Comroe's recommendation that "whenever scientists are asked to serve on such a panel to prepare a report on unclassified matters (such as the health of the nation), they should agree to serve only if it is clearly understood that their report—good or bad—will be available both to Congress and to the Executive Branch and, within a reasonable time, to the public."

It is apparently necessary to reaffirm unceasingly the need for openness in science and the dangers of secrecy. Openness is fundamental for scientific freedom and responsibility.

Protection of Individuals in Studies of Human Behavior

One necessary restriction on complete openness in science arises from the need to protect the integrity and privacy of individuals in studies of human behavior, and of communities in studies of social groups. Although we have already pointed this out in discussing the conditions for psychological research, it deserves further emphasis here. Physicians, in reporting on case studies, must respect the confidentiality of the physician-patient relationship. Inquirers into human sexual behavior have taken elaborate precautions to make sure that the identity of any person they interview remains unknown to any outside person. This is essential for the gathering of data, since many of the people interviewed would be subject to social embarrassment, and even to legal prosecution, if the information they provide were to be made public. Likewise, an anthropologist, interviewing people living under a government that they consider oppressive, must shield the identity of the people being interviewed and must convince them that they will be shielded. Otherwise they will not speak frankly, knowing that they might be subject to severe reprisals if they do.

Obviously, such requirements are essential for research on human beings and their attitudes and behavior. We emphasize the point here, since there have been occasional attacks on such research, on the ground that it is contrary to the open spirit of science. The restrictions are necessary, however, not only to protect confidentiality, but also to get honest answers from the people studied. There is the rare possibility, of course, that the investigator, using the confidentiality of his data as a shield, might falsify the results in a published report. There is more opportunity for successful falsification here than in most branches of science. Nevertheless, another independent investigator, though not interviewing just the same individuals, could conduct another study of a sample population, so that deliberate falsification of data would almost certainly come to light. It is, however, imperative that each investigator should clearly specify his sampling procedure, in order to permit others to check his results.

It is axiomatic in scientific observation and experiment that nothing can be detected or measured without affecting in some way the observed phenomena. An essential corollary to this is that all observation or experiment should be designed to minimize the disturbance introduced by the observer or experimenter, so that the true or "undisturbed" state of the system can be estimated. A second corollary is that meaningful data must include a conscious and careful estimate of the magnitude of the uncertainty introduced by the measurement process itself. An appreciation of these matters is important for issues of scientific freedom and responsibility. We distinguish here between the processes of science and the goals of science, the fundamental goals being the increase of knowledge and understanding.

The Committee believes that the vigor and integrity of science require that all areas of potential knowledge be open to inquiry; but the means of inquiry are open to challenge, particularly where life processes and human behavior are involved. A careful assessment of potential benefits weighed against probable risks is essential for the planning of observations or experiments that involve modification of human behavior or of the environment. The procedures proposed by the American Psychological Association, which we have discussed above, appear to assure adequate freedom for research together with scrupulous attention to the protection of the subjects of the research.

Technology and Innovation: Their Multiple and Complex Effects

We now turn from the problems of basic science to those of applied science and technology. Here the problems of freedom and responsibility are even more formidable and complex. Whatever the intentions of technological innovators, the results of innovation are always more complex than the innovators intended, and usually more complex than they could even imagine. These facts, under the present circumstances, create a social need to undertake the assessment, and the critical evaluation and control, of major technological innovations. The intended results of a new technology may be achieved, but they are commonly accompanied by a host of repercussions, most of which are unanticipated and often deleterious. To call them "side effects" may be misleading; the side effects sometimes dominate what was intended to be the primary effect.

The history of the use of DDT provides an example of such complexity (*35*). Its use brought a dramatic halt to a typhus epidemic in Naples during World War II; its initial success in destroying agricultural pests was spectacular. Only a few biological experts warned of possible hazards at that time. Only gradually did the troubles appear, as the insect pests developed resistance to DDT; as the natural enemies of the pests were destroyed by the pesticide itself in many cases; as the long persistence of DDT in the soil was discovered, and its progressive concentration in food chains led to the killing of certain birds and fishes in great numbers. In the Cañete Valley of Peru, for instance, the heavy use of chemical pesticides over a period of years proved disastrous; the crops declined, and the cost of pesticides rose, as more and more were used with increasingly less effective results (*36*). Finally, the local farmers banned synthetic organic chemical pesticides and worked out a control program that involved repopulating the valley with useful insects that the pesticides had destroyed. This was combined with the use of inorganic or natural organic pesticides, and with other control measures, which taken together gave successful results (*36*).

Even in the United States, where the damage from synthetic pesticides has been less drastic, unfavorable results have led to the banning of DDT for nearly all uses. Yet in countries where malaria is widespread, the spraying of house interiors with DDT has proved the most effective method available. In Sri Lanka (Ceylon), malaria had almost disappeared by 1963, when the DDT program was stopped. Then, in early 1968, there was an explosive increase of infections. Whereas in 1967 only 3,465 cases had been reported, the number rose in 1968 to 425,937, and for the first quarter of 1969 the number was 195,107, or nearly half of the total for the entire previous year (*37*). There were, of course, other causes for this dramatic change, besides the stopping of the DDT program; the World Health Organization attributed the rise in the incidence of malaria in large part to abnormally favorable breeding conditions for the *Anopheles* mosquito, and unusual human population movements that helped to spread the malaria parasite (*38*). Nevertheless, until we have something better, DDT appears to be an essential component of any major program of malaria control.

Other unanticipated complexities have arisen in programs for international development. Dams and irrigation schemes in many parts of Africa have vastly increased the incidence of bilharziasis (schistosomiasis), a disease that debilitates whole populations and for which there is as yet no effective cure (*39*). After the building of the first Aswan Dam in Egypt (1902) the incidence of schistosomiasis greatly increased, and with the construction of the High Dam it will probably increase further. There has been a progressive erosion of the Nile Delta, in the absence of the silt deposits that were formerly brought down by the annual flood, and which counterbalanced the forces of erosion (*40*). The sardine fisheries that once flourished outside the delta have been largely destroyed, although new fisheries may be established above the dam in Lake Nasser (*41*). It is very doubtful whether the various disturbing changes were taken seriously into account in a cost-benefit analysis before the dam was built.

The recent and current catastrophic droughts in sub-Saharan Africa (the Sahel) furnish a striking example of man-made technological damage. Climatic changes may have produced the droughts of recent years, but the droughts are probably no greater than many in the past that have not done comparable damage to the people or to the ecology of the region. The current disasters appear to be due largely to overgrazing and to the disruption of much of the traditional nomadic way of life. Particularly disastrous in their effects have been the thousands of deep boreholes, drilled by engineers to tap the vast reservoirs of underground water in that region. The resulting wells encouraged a great increase in the size of the cattle herds; pasture instead of water became the limiting factor on numbers of cattle. As pasture dried up in the drought, countless thousands of dead and dying cows were found clustered around the boreholes, and the surrounding land, for miles around, was ravaged by trampling and overgrazing. Thus these wells, constructed by men of good will and technical skill to bring more water to the people and cattle of the Sahel, became a major factor in intensifying a great human and natural disaster (*42, 43*).

The Necessity of Technology and of Its Assessment and Control

The illustrations we have just given could be multiplied a hundredfold. The wise use and control of technology is crucial for decent human life and indeed for survival. Let us add at once, however, that we are not among those who consider that there is something inherently evil about technology. The development of technology has been indispensable for the progress of all civilizations, and our own civilization cannot survive without new technologies. As one example, we need only mention the need for the development of new energy sources—solar and geothermal energy, and nuclear fusion, for instance—that may be less polluting than fossil fuels and less hazardous than nuclear fission.

Conversely, we reject the doctrine of the so-called "technological imperative"—the notion that, if something is technically possible, we must go ahead and do it. Thousands of projects may be technically feasible, including perhaps the destruction of all human life on earth. Even among those projects that appear attractive at first sight, careful appraisal may lead to the conclusion that some will do more harm than good. Congress drew such a conclusion in 1971 when it voted to cut off funds for the supersonic transport (SST) program, which a few years earlier Congress had been funding generously. Recently, Congress has recognized the more general class of problems that we have been discussing here by setting up the Office of Technology Assessment.

In summary, the development of new technologies is indispensable. The training and encouragement of gifted and imaginative technologists deserves a high priority among our national needs, but the multiple repercussions of new technology need to be critically evaluated before they are introduced, and constantly monitored after their introduction. Many schemes that are technically brilliant must be rejected because their wider impact would, on the whole, be more damaging than beneficial. In some cases it might be preferable, in the eyes of many good judges, not to carry a project from the stage of research even into preliminary development, lest pressures would then arise that would lead to its full development.

Planning and the Unexpected in Applied Science: Problems of Responsibility and Choice

Many of the greatest discoveries in basic science—x-rays, radioactivity, nuclear fission, nuclear magnetic resonance, lasers, to name a few—were quite unexpected. No committee planning science budgets for the future would have anticipated them before they occurred. Hence wise administrators have recognized that scientists doing basic research need wide latitude in their choice of problems and in the way they operate. Even in basic science, however, most research is not on this highly original level; it involves exploitation of leads already given by others.

In applied science and development most researchers have little freedom to pick the subject of their research. Research programs in most governmental and nearly all industrial laboratories are planned with practical ends in view. The aims of the research workers are, along broad lines, laid down for them by their superiors (*44*). They may be under general instructions to find a better and more selective pesticide, to hold some damaging insect in check; to improve the accuracy of an intercontinental missile; to prepare an improved plastic; to develop a new kind of tear gas for the dispersal of crowds or for use in war; to develop more efficient and less noisy aircraft engines; or any one of ten thousand other problems. Within the limits of such a general aim, the researchers may be allowed wide latitude in the choice of approaches to the problem, depending on their individual ability and character, and on the general policy of the laboratory. An effective solution for such an assigned problem may involve a deep knowledge of basic science and close familiarity with the latest fundamental developments in the field. The gap between basic and applied science is certainly narrowing; new fundamental discoveries, such as lasers, are converted into marketable products with a speed that would have been inconceivable half a century ago. The solution of applied problems, in science and engineering, can often involve daring leaps of the imagination. The satisfactions to be derived from this type of work are, for many gifted researchers, at least as great and for some often greater than for work in basic science. We have no sympathy with the attitude of many academic scientists, who look down on workers in applied science and engineering as having inferior status to the "pure" scientists. The satisfactions of the workers in applied science can be very great, and much of the future of the world is in their hands; we need the best such people that we can get (*45*).

In a thoughtful article, Steven and Hilary Rose have raised the question: "Can Science Be Neutral?" (*46*). Their answer is an emphatic "no," and in the domain that we are here considering they are certainly right. They illustrate their point in terms of the development of the "super tear gas," commonly known as CS, which was widely used in Vietnam to drive soldiers and civilians out of shelters into the open and also used to quell disturbances in Northern Ireland. They show that CS was developed as a matter of clear and conscious choice, under the direction of scientific administrators who knew what they wanted and got it. The process involved use of basic chemical researches done a generation earlier, but these later developments were deliberately aimed at a known objective.

The decisions of high-level scientific administrators naturally reflect in general the policies of their governments. Many scientists and others have opposed the use of CS, particularly its use in war; but they could not bring their influence to bear on policy decisions. The development of CS is only one small aspect of the onrushing technology of the arms race, which has acquired a momentum of its own. Recognizing the unprecedented perils to which mankind is subject, not only from nuclear weapons but also from chemical and biological weapons, many conscientious scientists would categorically refuse to give any assistance to further weapons development. This is, for instance, the position of the Society for Social Responsibility in Science. We profoundly respect the position of such conscientious objectors; but we also note that they are far outnumbered by the scientists who work, most of them naturally in secret, in the development of newer and more deadly weapons, and who for the most part would hold that in doing so they are performing a patriotic service in the interest of the national security.

The makers of policy in the United States, the Soviet Union, and other countries know well that their countries are subject to mutual destruction if nuclear war breaks out. No one can hope to gain; the threat of unprecedented catastrophe hangs over all. Awareness of this fact has imposed the need to resolve conflicts without resort to nuclear war. These fears, however, have not inhibited but have on the whole been used to accelerate the arms race, as the government and military leaders in each nation, fearing (or claiming fear) that their rivals will develop some new weapon to upset the balance of deterrence, press ahead with weapons developments that stir the fears of their rivals in turn, and thereby encourage them, or give them the excuse, to do likewise. Thus each side is psychologically trapped by fear of the other, and the arms race spirals upward. The recent failure of the SALT talks, and of the Soviet-American summit meetings of 1974, to produce significant steps toward arms control and reduction in armaments, are alarming and disheartening (*47*). Only a very small group of scientists has really attempted to grapple with these awesome problems; they include those involved in the Pugwash Conferences, some very active members of the Federation of American Scientists, and some contributors to such journals as *The Bulletin of the Atomic Scientists*. Matters of life and death for society, such as these, are of course everybody's business, and especially the business of politicians; but they can hardly be solved without a deep involvement by many scientists, far more than the number of individuals who are now working on these problems.

Problems of Responsibility in Medicine and Public Health

Physicians and public health workers are clearly responsible to the individuals and the community whose health they are under obligation to protect. Serious conflicts may arise when the physician works for a commercial firm whose profits may be threatened by the added cost of steps required to protect the health of the workers. In recent books Paul Brodeur (*48*) and Rachel Scott (*49*) have described the medical problems of workers exposed to asbestos particles, to beryllium dust, and to other industrial hazards. Many of these workers develop cancer, or fatal respiratory diseases, over the course of 10 to 20 years; the risk is high. Yet Brodeur's account shows that the company physicians involved sometimes join their employers in opposing the introduction of more rigorous standards of health protection. This appears to be a clear abdication of the responsibility of physicians (or public health workers) to protect the life and health of the people under their supervision.

Dr. Alice Hamilton, one of the great leaders in industrial medicine during the first third of the 20th century, in her autobiography (*50*) described in vivid detail her battles to protect workers exposed to dangerous chemicals. Thanks to her efforts and those of many others, the laws for the protection of workers have been greatly strengthened; but the battles go on, as new chemicals are introduced and new hazards are discovered, and as some powerful industrial managers resist the introduction of expensive protective measures.

Clearly a doctor who is paid by a commercial enterprise will find it very difficult to act in opposition to the policy of the company. Apart from questions of loyalty to his employer, he may have little choice except to resign if he fights the management. If he is prepared to fight, he can often force an investigation by a governmental agency concerned with public health; but, even if he wins his fight, he will probably be fired or forced to resign in the end. To strengthen his ability to defend the public health, he needs support from a powerful professional organization that is dedicated to the same objective. Of the role of such organizations we speak later in more detail. If he is an employee not of the company but of an independent inspecting agency, his position will obviously be much stronger. Here we reaffirm the principle that the promotion of an activity, and the regulation of that activity in the interest of public health and safety, are two distinct functions, and should ordinarily be vested in separate agencies. The regulatory agency must stand on its own feet and must be safeguarded from pressure by the promoters, whether the latter be a business firm or a governmental agency. Regulatory agencies, in the course of time, frequently tend to become subservient to the organizations that they are supposed to regulate. Constant vigilance by the concerned public is necessary to prevent this.

Chapter Three

The Second Charge

Conflicts Involving Scientific Freedom and Responsibility; Means for Achieving "the Objective and Impartial Study of These Problems"

We now turn to the second item in the charge to the Committee, which calls for us "to develop suitable criteria and procedures for the objective and impartial study of these problems." We have already provided a number of illustrations of the conflicts that involve issues of scientific freedom and responsibility; we shall consider more in the following discussion of criteria and procedures. To ensure that the procedures are objective and impartial is a formidable problem, for we are dealing with profoundly controversial issues, in which one or more scientists may be in sharp dispute with a powerful industry or government agency, and where often one group of scientists takes issue with another, as with the debate over the antiballistic missile a few years ago. These debates call into question the very meaning of the word "objectivity"; one person's objectivity appears as bias to another.

An important example of a conflict of this sort, and at least its partial resolution by a presumably objective body, is the dispute over the hazards of ionizing radiations that began about 6 years ago. Drs. John W. Gofman and Arthur Tamplin, who were working for the Atomic Energy Commission (AEC) at its laboratories in Livermore, California, claimed that the official standards for permissible exposure to nuclear radiations were far too tolerant, and that exposure of the population to radiation at the permitted level would result in some 30,000 additional cases of cancer in the United States each year. They publicized their views in lectures, magazine articles, testimony before congressional committees, and in other ways. Authorities in the AEC hotly disputed their claims, and several AEC scientists wrote articles intended to disprove them. Gofman and Tamplin reported that the AEC, in retaliation, was cutting their laboratory budgets and drastically reducing the number of co-workers allotted to them. Indeed, both of them, within about 2 years, left the AEC and found employment elsewhere. The AEC, however, took note of the protests by sharply lowering the acceptable radiation levels in the facilities operated under their auspices. In the endeavor to resolve the issue, the National Academy of Sciences appointed a committee of experts to review the evidence on the hazards of ionizing radiation. Their report, commonly known as the BEIR Report (*51*), stands at present as probably the most thorough and objective discussion of this controversial issue. The report is in agreement with the view of Gofman and Tamplin that there is no threshold exposure level below which radiation will not induce cancer or genetic damage; but the quantitative estimates of damage in the BEIR Report are a good deal lower than the estimates of Gofman and Tamplin.

In this very brief discussion, we cannot do justice to the history of the case or the complexity of the issues. Gofman and Tamplin lost their jobs with the AEC; although they were not technically fired, it seems fairly clear that the AEC made their position sufficiently uncomfortable to force them out. Obviously, the AEC considered them unnecessarily obstreperous, whereas they were convinced that they had to be obstreperous in order to be heard. The scientific issues received, in the BEIR Report, about as objective an examination as such a difficult and controversial problem could be expected to receive at the present time.

Other controversial issues are now being increasingly referred to the National Academy of Sciences by Congress, by the Executive Branch, and sometimes by other sources. The Academy has the right to undertake studies of important issues on its own initiative, even in the absence of a request from the government. Clearly it can call upon authoritative bodies of experts, in very diverse fields, to deal with many difficult issues. Critics have charged some Academy committees with bias, a charge that has sometimes been justified in the past. However, the Academy's Report Review Committee, which sees to the careful critical examination of all major reports, has been a powerful influence for detecting and correcting bias in various committee reports. The Academy will certainly continue to play a major role in resolving some of the disputes relating to scientific freedom and responsibility.

On the governmental level, the National Environmental Policy Act has provided, through the requirement for Environmental Impact Statements, a mechanism for critical and balanced examination of projects that will have significant ecological repercussions. In a number of cases, ill-conceived projects have been blocked or drastically modified by the requirement for the preparation and publication of such statements, and by the subsequent public hearings. However, the relative power of the contending parties is often grossly unequal; a powerful government agency, with strong political and industrial backing, can often prevail in its purposes, even though the weight of expert scientific testimony may be on the other side. Active private watchdog organizations, such as the Natural Resources Defense Council and the Environmental Defense Fund, can often help to redress the balance by calling on scientific and legal experts, but their resources and funds are limited.

Organizations of private citizens can play an important role in obtaining balanced consideration of important controversial issues. Thus, the Federation of American Scientists (FAS) has, in the last few years, furnished a critical examination of some of the budgetary demands and the research and development proposals of the Department of Defense. Experts such as G. B. Kistiakowsky and Herbert F. York, who have great knowledge of military problems but are quite independent of the Defense Department, have given important testimony under the auspices of the FAS, which has provided important perspectives in considering the military budget.

The Scientists' Institute for Public Information (SIPI) in New York, which publishes the magazine *Environment* (*St. Louis*), is an important private organization that aims to provide the general public with essential and reliable facts, on as impartial a basis as possible, concerning the environment, energy, and other important issues in which some scientific background is essential for intelligent judgment by the public. SIPI endeavors to be informative, without taking sides on controversial issues. Such a dispassionate attitude, of course, is difficult to maintain consistently; a writer may believe that he is being objective, while his critics are equally convinced that he is selecting data to prove a thesis. Certainly, however, SIPI has performed, and continues to perform, valuable service in setting forth essential and reliable information that is necessary for making sound decisions on many public issues that involve science and technology.

Private *ad hoc* organizations, formed to deal with particular issues, can play an important role. Generally, they are not impartial but are formed to support one side in a dispute. However, if that side has not received an adequate hearing, the activity of such a group may serve to present to the public a more balanced view of both sides of the case than would otherwise have been possible. An example is provided by the history of the supersonic transport (SST) program. About 1960, the British and French governments (in the Concorde project), Russia, and a little later the United States, all initiated programs for building commercial SST's. The scheme was enthusiastically promoted in all these countries; the promoters hailed the prospect of speeding long-distance travel and enhancing national prestige with a dramatically new type of plane. The first and most important critic of the SST was the Swedish aeronautical expert Bo Lundberg. In a series of articles, he emphasized the adverse effects of the sonic boom, pointed out various safety hazards of the plane, and questioned whether it could be economically competitive with other forms of air travel. In the United States, the physicist W. A. Shurcliff took up Lundberg's arguments and founded the Citizens' League against the Sonic Boom, which attained a membership of several thousand. Working with voluntary private contributions from members of the League, Shurcliff powerfully publicized the case against the SST. Later, a coalition of environmentally concerned organizations joined the anti-SST forces and lobbied effectively in Washington. Finally, in 1971, their efforts were so successful that Congress voted to cut off all government funds in support of the SST program—a result that would have seemed inconceivable 2 or 3 years before (*52*).

In this instance, there was another important factor in the Congressional decision to stop supporting the SST program. Late in 1970, a group of eminent economists testified before a Congressional Committee that the SST would probably be an economic failure. Thus, while the environmentalists pointed out that a large SST program would do various kinds of environmental damage, the economists pointed out that it would also lose money. The combined attack proved fatal to the project, at least for the time being. Schemes that do environmental damage but can make large amounts of money for their promoters are not so easy to stop.

We should call attention here to one aspect of the SST controversy that emerged only after the Congressional vote of 1971. H. Johnston, a distinguished chemist at the University of California at

Berkeley, proposed that the nitrogen oxides that would be liberated in the stratosphere if a large fleet of SST's went into operation would seriously diminish the ozone content of the stratosphere and thereby permit dangerously large amounts of ultraviolet light to reach the surface of the earth. This would lead to increased skin cancer and other forms of biological damage (*53*). The calculations involved are extremely complex, involving the kinetics of a large number of chemical reactions as well as the physics of the upper atmosphere; and Johnston's numerical results have been challenged by other investigators (*54*). Obviously, we have no competence to offer a technical judgment in this controversy. We can affirm, however, that the raising of such questions by well-qualified scientists, and their systematic study, is an excellent example of scientific responsibility, of the sort that will be increasingly needed in future. We must known in advance, if possible, the many repercussions of new technological innovations, both in order to prevent or minimize environmental damage and in order to avoid wasting money or schemes that ultimately prove to be undesirable.

Criteria and Procedures in Issues of Scientific Freedom and Responsibility: The Role of Professional Societies. The Case of the BART Engineers as an Example

The brief case histories we have provided above illustrate many of the issues that arise concerning scientific freedom and responsibility. With this background we now consider criteria and procedures for resolving such issues.

We approach the problem of establishing criteria by asking questions rather than by offering answers. Some of the questions are likely to be: How will the proposed decision or procedure affect human health and safety, and the general quality and amenities of life, for the people affected? Will a decision to require a drastic cleanup of operating conditions in a certain industry cause a shutdown of the industry, with loss of jobs and of the products of the industry? How are such risks to be balanced against considerations of safety and health? Are some individuals being exploited by being exposed to the hazards of disease and premature death to enhance the profits of the manufacturer and make a cheaper product? What are the possible large-scale environmental effects of the present operations and of the proposed changes? These effects may be subtle, complex, and difficult to perceive in advance. Have possible future effects been carefully considered, as with the Freon compounds in aerosol spray cans, which in a decade or two might (but again perhaps might not) destroy much of the ozone in the stratosphere, with highly deleterious results from increased ultraviolet radiation? We should also ask another question, important both for scientific integrity and public confidence: Is somebody lying, or deliberately trying to deceive the public? It would be easy to raise more questions, but these are sufficient to illustrate the problem of formulating criteria.

Clearly the most frequent sources of conflict involve economic interests, on the one hand, and values of health and environmental quality, on the other. This is dramatically illustrated by the recent lawsuit against the Reserve Mining Company, which has for years been dumping tens of thousands of tons of taconite particles each day into Lake Superior near Duluth, as a by-product of its mining of iron ore. Taconite is a form of asbestos, and asbestos is carcinogenic if inhaled over long periods. Judge Lord, who presided over the case, finally ordered all operations shut down until a suitable way of disposing of the taconite on land was worked out. Almost immediately, however, a Court of Appeals overruled him and allowed the company to continue operations for a period of up to 5 years, while noting at the same time that a "monumental environmental mistake" had been made in allowing the procedure in the first place (55).

We make no claim to have definite answers to all these difficult problems. Our own bias would favor resolving difficult issues in favor of human health and environmental quality, as against primarily economic considerations; but we would most urgently plead for increased foresight to anticipate such dangers at an early stage, before large-scale industrial development occurs, so that these grim conflicts between opposing interests may be avoided as far as possible.

We turn now from criteria to procedures for the resolution of such conflicts. Many scientific and engineering societies have developed codes of ethics, relating to the responsibility of employers and to the professional and personal conduct of scientific and technical employees. A highly articulate expression of such concerns is to be found in a statement on "Employment Guidelines," which has now been adopted by at least 20 engineering and scientific societies. For the most part, it is concerned with the general principles that should govern relations between employers and employees, but it also contains the significant statement: "The professional employee should have due regard for the safety, life and health of the public and fellow employees in all work for which he/she is responsible. Where the technical adequacy of a process or product is involved, he/she should protect the public and his/her employer by withholding of plans that do not meet accepted professional standards and by presenting clearly the consequences to be expected if his/her professional judgment is not followed" [(*56*), p. 59].

The formulation of such a declaration is a significant event. How much it means depends, of course, on the effectiveness with which it is applied. Moreover, these guidelines, like most such codes of ethics that we have seen, lack a very important ingredient, namely, a provision for the arbitration of disputes. The protection of individuals from arbitrary action by authority is deeply ingrained in English common law, and the U.S. Constitution provides that "no person shall . . . be deprived of life, liberty, or property without due process of law." We believe that some form of due process should be an essential part of any employer-employee agreement or contract, to protect the employee from arbitrary action by the employer, allegedly based on professional or personal misconduct. A minimum requirement for such due process would involve a hearing by a board, including independent members, with the right of appeal to some reasonably neutral but professionally qualified higher authority. Codes of professional ethics are likely to be ineffective unless some type of due process is provided for the resolution of disputes. Without this, scientific freedom is likely to be abridged. We therefore strongly recommend that all employment contracts involving scientific or professional employees include such provisions for the review of disputes through hearing and appeal processes. Provision for neutral or third-party participation is important, particularly when issues of public interest are involved.

An important case, in which these principles are certainly involved, arose in connection with the building of the San Francisco Bay Area Rapid Transit (BART) system (*57*). A major feature of BART was to be the Automated Train Control (ATC) system. A contract for the development of this system was awarded in 1967 to the Westinghouse Electric Corporation. Holger Hjortsvang, a systems engineer in the BART maintenance section since 1966, became increasingly concerned about defects in the ATC system and expressed these concerns in a series of memoranda to his superiors, beginning as early as April 1969. There was no significant response from the management. Max Blankenzee, a programmer analyst working with Hjortsvang, had a similar experience. His warnings drew only vague verbal responses and advice not to be a "troublemaker." Robert Bruder, an electrical engineer in the construction section of BART, grew increasingly disturbed about the "unprofessional" manner in which the installation and testing of the ATC system was being carried on. Again the management disregarded his repeatedly expressed concern.

Late in 1971 the three engineers decided that, in the public interest, they must take further steps to have their concerns considered seriously. They went to Daniel Helix, a member of the BART Board of Directors. This led to a public hearing by the BART Board in February 1972, in which the criticisms of the ATC system were presented, and in which Westinghouse and associated engineering firms defended their operations. The Board voted 10 to 2 in support of the management, in effect rejecting the criticisms offered by Hjortsvang, Blankenzee, and Bruder. In early March, the managers then told the three that they could choose between resigning and being fired. On refusing to resign, they were summarily dismissed.

Subsequent events of course vindicated their concern. The ATC system failed on several occasions; on 2 October 1972, a BART train overran the station at Fremont, and several passengers were injured. Other such failures occurred so often that it became necessary to control the trains in the traditional manner.

After the firing of the three engineers, the California Society of Professional Engineers (CSPE) initiated an inquiry; BART's top management refused to meet with CSPE representatives, or indeed to offer any explanation to anyone. CSPE then undertook a full investigation of the firings, and of the conduct of the three engineers. President Jones of CSPE stated that he and other CSPE members were "convinced that the three engineers acted in the best interest of the public welfare" and also that "a large volume of most distressing information on the employment practices of BART, and on its apparent disregard for public safety, has been gathered." The California State Legislature then launched an investigation, which resulted in a report (the "Post Report") which essentially confirmed the warnings of the three engineers. CSPE took some tentative steps toward a court action on behalf of the three but did not follow through on this. The three engineers themselves launched a suit against BART for damages totaling $885,000. The outcome of this we have not yet been able to learn.

We have told this story in some detail, since it is a particularly dramatic illustration of the problems facing scientists and technologists who attempt to defend the public interest in the face of powerful opposition and obstruction. The supporting action of a professional society—CSPE in this case—helped to publicize the issue and to bring it before the Legislature; the plan for CSPE to file a legal suit on behalf of the dismissed engineers, however, failed to materialize.

How active can, and should, professional societies be in actively fighting on behalf of their members who are attempting to defend the public interest? Most such societies have in the past remained aloof from conflicts of this sort, and have often taken the attitude that the purity of their devotion to the advancement of their respective sciences would somehow be contaminated if they entered the public arena to contest such issues. We believe that such attitudes are no longer appropriate. The scientific community can no longer remain apart from the conflicts of our time, where so many technological decisions are being made that vitally affect the well-being of society. We are not proposing that professional societies should take public stands on large general political issues, such as the legitimacy of the Vietnam War; individual members of the societies, when their concern is aroused, should deal with these matters by other mechanisms. However, in matters directly related to the professional competence of members of the society, where the public interest is clearly involved, we believe that the societies can and should play a much more active role than in the past. They can deal with such issues by setting up committees of inquiry, in cases where a serious violation of scientific responsibility is suspected; by publicizing the results of the inquiry in professional journals, and, if necessary, in the more popular journals and in the news media; and by calling the matter to the attention of governmental bodies, as with the California Legislature in the BART case. They can on occasion launch lawsuits on behalf of members who have apparently suffered injustice when acting on behalf of the public interest.

In stating this, which is our major new proposal for dealing with "the objective and impartial study of these problems," we are aware of the difficulties that the proposal will face. The most serious problems are those of time and money. Most professional societies have limited funds; many operate more or less on a shoestring. They keep members' dues fairly small; otherwise members drop out, particularly in times of economic hardship. The fighting of difficult cases, on behalf of members involved in controversy, can be a very expensive business, especially if the case goes into the courts. In any case, it would require that responsible scientists spend time in serving on hearing panels, studying large bodies of evidence, and preparing reports; this would involve a substantial sacrifice of precious time for them.

When a professional society does fight for the rights of its members, it is more likely to be concerned with defending their status and pay than to be acting primarily on behalf of the public interest as its primary motive. The impetus to take actions of the latter sort is likely to be much less strong than the desire to provide direct help to members of one's own professional group.

These are powerful obstacles to our proposals, but they are not insuperable. Societies that share common interests, but which may be individually too weak financially to support such activities, may band together in groups to finance the necessary operations. There are increasing pressures upon scientists, engineers, and other members of the scientific community to face these public issues and deal with them effectively. These pressures come both from the public and from within the ranks of the scientists themselves. We have spoken, earlier in this report, of the mistrust and hostility toward science that is manifest in many quarters; one reflection of this attitude is the decline in government support of science in recent years. Such hostility will almost certainly grow unless scientists exhibit greater concern for preventing the misuse of science and technology. As this becomes clearer, it will become easier for the professional societies to obtain additional funds to carry the expenses of lawsuits, hearing panels, and other activities undertaken in the defense of the public interest. Whether government funds could or should be available for such purposes is open to question; but it is likely that some of the major private foundations, either those now in existence or those yet to be created, will see the urgency of supporting such public service activities. The need for these activities may also lead to the creation of other social mechanisms for dealing with these problems, of a sort that we cannot now foresee. We look to increased activity of the professional societies as the most hopeful approach to the problem in the immediate future.

Chapter Four
The Third Charge

On the Role of the AAAS in the Defense of Scientific Freedom and Responsibility

We must now consider the third charge from the Association to this Committee: "To recommend mechanisms to enable the Association to review specific instances in which scientific freedom is alleged to have been abridged or otherwise endangered, or responsible scientific conduct is alleged to have been violated." The domain of the AAAS is very broad; it includes all the sciences, and a large number of professional societies belong to it as affiliates. Hence, by its nature it cannot undertake the role that we have envisaged for the professional societies in the preceding section. The number of possible appeals to it, if the Association were to undertake the responsibility of reviewing particular instances of alleged threats to scientific freedom and responsibility, would be immense. If it were to agree to handle some such cases, it would have to be rigorously selective and deal only with cases that were at once so important, and so broad in scope, that they would fall outside the domain of any individual society.

Because of its multiple affiliations with the professional societies, the AAAS can play an important part in coordinating many of the activities of the societies. A number of the societies, for example, are now in the process of formulating, or revising, codes of ethics for their members, which will involve policies for dealing with issues such as we have just been discussing. The AAAS can help to provide an exchange of information on these matters among the different societies, and thereby promote a more unified approach to these complex issues.

The AAAS is already playing a significant role in dealing with alleged abridgments of scientific freedom, and alleged violations of responsible scientific conduct, through active discussion of such matters in *Science*, chiefly in the "News and Comment" section but also in some of the leading articles and in the "Letters" section. Since *Science* is so widely read, both inside and outside the scientific community, this is one of the most important channels now available for bringing such problems to the attention of the public. To focus public attention on such problems, of course, is not to resolve them, but it is an essential step toward such resolution. We recommend that *Science*, without any drastic change of its present editorial policies, enhance its coverage of such matters, particularly by inviting distinguished scientists who are well informed on some of these controversial issues to set them forth in its pages.

In some cases, it will be desirable to present two or more articles by different authors, expressing more or less contradictory points of view. In scientific controversies, it should not be necessary for the champions of different views to operate like adversaries in a court of law; in a scientific controversy, the opposing sides can presumably find a large area of agreement about scientific facts that are not in dispute. When eminent scientists disagree vehemently about issues that involve a large scientific and technological component, the public becomes confused and is likely to think that scientists are not really scientific after all. The real disagreements, however, usually hinge not on the scientific facts but on the relative weight given to different *kinds* of scientific facts, and on extrascientific issues involving political judgment and broad general perspectives on human nature and human motives. These factors always enter into the practical decisions that have to be made in the applications of science and technology. When a scientist or technologist states a case for action of a certain sort on such an issue, it is important to make clear the general presuppositions on which the argument is based. He may, of course, be unconscious of these presuppositions; if so, it should be a matter of editorial policy to bring them to the surface, before an article of this sort is published in *Science*. This is important both for the rational discussion of the issues involved and for the maintenance of public confidence in the honesty and objectivity of scientists.

There is a similar opportunity for the effective presentation of these controversial issues at the annual meetings of the AAAS, and indeed the Association has performed an important service in presenting symposia on some of these issues at several of its meetings.

What we have already said in the preceding paragraph applies here with little change.

On rare occasions, the AAAS may properly become involved with issues of broad and fundamental character that may lie beyond the scope of a more specialized society. An example of such involvement was the decision of the Association to conduct an investigation of the effects of defoliants and herbicides in Vietnam, and to send out a team of scientific and medical experts to make detailed observations in that country. This was an important attempt to obtain scientific evidence on an issue that had stirred passionate controversy, and indeed continues to do so. This investigation serves as an example of the sort of major problem, involving scientific responsibility, in which the AAAS should take the initiative (*58*). Problems of similar importance will be rare, and their character may vary greatly from case to case. Hence, we make no attempt to suggest patterns of procedure for dealing with cases that may arise in future; the choice of procedure must be determined by the nature of the problem.

We note that the National Academy of Sciences, at the request of Congress, later undertook a study of defoliation and herbicides in Vietnam. This was financed by the Department of Defense, on a more generous scale than was possible for the AAAS study. The report of the National Academy Committee has now been published (*59*); in large measure, the findings of the two reports are reasonably concordant, but some discrepancies between the findings in the two reports have given rise to controversy (*60*). It seems clear that the investigation sponsored by the AAAS was in no way rendered superfluous by the more comprehensive National Academy study. Not only did it provide important information at an earlier stage, but the disagreements on certain points between the two studies will lead to further investigation to resolve the differences. Such further work is at present rendered nearly impossible by the continuing war in Vietnam, but, when conditions become more peaceful, it will be possible for scientists to make observations on the ground in the Vietnamese forests and settle disputed questions by direct observation.

These events do raise a more general question: what should be the relation between the AAAS and the National Academy of Sciences in dealing with problems of the general sort that we have been discussing here? The National Academy has special prestige and a special relation to the U.S. Government. It can speak on many issues with far more authority than the AAAS, and it can generally command much greater financial support for its investigations. However, the AAAS, with its much broader membership, is more widely representative of American science in general, and the fact that it has no governmental ties gives it in many cases a greater freedom of action. The relation between the two organizations should be worked out in future with careful consideration of their appropriate roles.

The problems we have been considering in this report will certainly continue and will probably become more numerous and more acute in the years to come. The AAAS will have a continuing concern for them. We suggest that, not more than 5 years hence, the Association should reexamine the whole problem, perhaps by setting up a committee similar to this one, to see where we stand at that time. Alternatively or additionally, it might set up a committee to receive complaints concerning violations of scientific freedom and responsibility, and refer them, when possible and desirable, to appropriate bodies for further study and possible action. Such a committee should not itself serve as a judicial body; its functions should be to refer complaints elsewhere for possible action and to analyze the information received, with a view to possible recommendations concerning future policy initiatives by the AAAS in the light of this information. The terms of reference of such a committee would have to be very carefully drawn, to prevent it from being overwhelmed by a mass of unmanageable complaints.

Chapter Five

Some Closing Words

In our time, scientists have seen an enormous increase in the influence of their discoveries on the conditions of life and the world's prospects. For hundreds of millions of people the fruits of science have yielded a vast increase in comfort, health, and mobility. Far larger numbers of people still live in poverty and often on the edge of starvation. Looking back on past history, we see that large areas of the earth, once fertile and the homes of great civilizations, have become far less hospitable to life and today are often deserts or semideserts. This has happened in the Middle East and the Mediterranean Basin, and in parts of East Asia, Africa, Latin America, and the United States. Much of this decline is due to damage from earlier technology. With the vastly more powerful technology available today, we can do more damage, and do it far more quickly, than was possible for our ancestors with their more primitive techniques. On the other hand, we can begin to foresee, much more clearly than they did, the multiple and complex consequences of the actions that we take; potentially we can use technology with more wisdom than ever before. The endeavor to achieve that wisdom is everybody's business; but the responsibility of scientists is obviously much greater than that of nonscientists. We are not urging that all scientists should work on problems related to the applications of science; there is an enduring need for the devoted researcher-teacher, who loves science for its inherent beauty and fascination and seeks to impart that love and understanding to others. Yet increasingly, for many of us, it is impossible to feel the same delighted fascination with science that we once did, without also being deeply concerned with the uses and misuses of science that will largely determine the future of mankind. It is in this area that scientists need both the freedom to speak out and the responsibility to speak and influence policy, on the basis of all the knowledge and wisdom they can muster. It is particularly the obligation of scientists to take the long view, to consider how the actions we are taking now will affect the condition of the earth, and of mankind, a hundred or ten thousand years from now. Politicians and industrialists must necessarily think in shorter terms; scientists too must deal with today's problems, but they must also look far ahead, as best they can. It is in the endeavor to make a small contribution to this end that we have submitted this report.

References

1. Environmental Health Perspectives, Experimental Issue No. 5: *Perspective on Chlorinated Dibenzodioxins and Dibenzofurans* [Department of Health, Education and Welfare Publication No. (NIH) 74-218, National Institute of Environmental Health Sciences, Box 12233, Research Triangle Park, North Carolina, Sept. 1973; iv + 313 pp.]. Several articles deal with the toxicity of the dioxins. See particularly table 4 in the article by R. Baughman and M. S. Meselson, pp. 27–35, which summarizes the results of earlier workers.
2. J. A. Moore, "Creationism in California," *Daedalus* 103 (No. 3), 173–189 (1974).
3. V. Earle, Ed., *On Academic Freedom* (American Enterprise Institute, Washington, D.C., 1971; 52 pp.).
4. R. K. Merton, *The Sociology of Science* (a collection of papers) (Univ. of Chicago Press, Chicago, 1973; xviii + 605 pp.).
5. M. Polanyi, *Personal Knowledge: Towards a Post-Critical Philosophy* (Harper Torchbooks, New York, 1964; xvi + 428 pp.).
6. J. M. Ziman, *Public Knowledge* (Cambridge Univ. Press, London, 1968; 152 pp.).
7. W. O. Hagstrom, *The Scientific Community* (Basic Books Inc., New York, 1965; xiv + 304 pp.).
8. J. R. Ravetz, *Scientific Knowledge and Its Social Problems* (Clarendon Press, Oxford, 1971; xii + 449 pp.).
9. B. Barber, *Science and the Social Order* (George Allen and Unwin, Ltd., London, 1953; 288 pp.).
10. D. S. Greenberg, *The Politics of Pure Science* (World Publishing Company, New York, 1967; 303 pp.).

10a. J. Primack and F. von Hippel, *Advice and Dissent: Scientists in the Political Arena* (Basic Books Inc., New York, 1974; xii + 299 pp.).

11. F. R. Jevons, *Science Observed: Science as a Social and Intellectual Activity* (George Allen and Unwin, Ltd., London, 1973; 186 pp.).
12. K. R. (Sir Karl) Popper, *The Logic of Scientific Discovery* (Basic Books Inc., New York, 1959; 480 pp.). See also K. R. Popper, *Conjectures and Refutations: The Growth of Scientific Knowledge* (Harper Torchbooks, New York, 1965; xiv + 417 pp.).
13. P. B. Medawar, *Induction and Intuition in Scientific Thought* (American Philosophical Society, Philadelphia, 1969; x + 62 pp.).
14. For an illuminating study of the referee system, see H. Zuckerman and R. K. Merton, *Minerva* 9, 66–100 (1971); reprinted in (4), pp. 460–496.
15. A. J. Ihde, *The Development of Modern Chemistry* (Harper and Row, New York, 1964; xii + 851 pp.), p. 401.
16. J. Gaston [*Originality and Competition in Science* (Univ. of Chicago Press, Chicago, 1973; xx + 210 pp.] includes much interesting material on competition among scientists and the theft of scientific ideas. His study deals with the community of high-energy physicists in Great Britain, but similar situations would almost certainly be found among other groups of scientists and in other countries.
17. Z. A. Medvedev, *The Medvedev Papers* (Macmillan, London, 1971; xiv + 471 pp.).

18. J. T. Edsall, "Government and the Freedom of Science," *Science* **121**, 615–619 (1955).
19. Report of the Committee on Loyalty in Relation to Government Support of Unclassified Research, "Loyalty and Research," *Science* **123**, 660–662 (1956).
20. B. Nelson, *Science* **164**, 1499–1504 (1969); *ibid.* **165**, 269–271 (1969); *ibid.* **166**, 357 (1969); *ibid.*, p. 487.
21. This attitude is eloquently expressed by T. Roszak, "The Monster and the Titan: Science, Knowledge and Gnosis," *Daedalus* **103** (No. 3), 17–32 (summer 1974). See also G. Holton, "On Being Caught between Dionysians and Apollonians," *ibid.*, pp. 65–81. For an illuminating discussion of the anti-science movement and its historical background, see *Civilization and Science: In Conflict or Collaboration?* [Ciba Foundation Symposium 1 (New Series), Elsevier Excerpta Medica, North-Holland, Amsterdam, 1973], especially the articles by P. B. and J. S. Medawar, by S. Toulmin, by E. Shils, and the discussion that follows.
22. P. Berg *et al.*, *Science* **185**, 303 (1974); N. Wade, *ibid.*, pp. 332–334.
22a. The moratorium on these genetic experiments was subjected to intense critical review at an international conference of experts in February 1975 at the Asilomar Conference Center in Pacific Grove, California. Paul Berg served as general chairman. After intense and sometimes impassioned discussion, the conference approved a statement lifting the moratorium on the great majority of the proposed categories of experiments, but replacing it by extremely stringent controls. The stringency of the required precautions is to increase with the estimated degree of risk from the experiments. Indeed some of these experiments are supposed to be done only with mutant strains of bacteria that cannot survive outside the laboratory—strains that do not yet exist and that must be developed by experiment. Thus for practical purposes the moratorium is being continued, for the time being, for such experiments. For further details regarding this important conference see N. Wade, *Science* **187**, 931–935 (1975).
23. A. R. Jensen, *Educability and Group Differences* (Harper and Row, New York, 1973; 407 pp.).
24. T. Dobzhansky, *Genetic Diversity and Human Equality* (Basic Books Inc., New York, 1973; xii + 129 pp.). See also A. R. Jensen, *Behavior Genetics* **4**, 1–28 (1974).
25. H. Brooks, "Scientific Response to Public Concerns about the Atmosphere," in *Modifying the Weather: A Social Assessment*, W. R. D. Sewell *et al.*, Eds. (University of Victoria, Victoria, British Columbia, 1973), pp. 83–96. The quoted passage is on p. 86. The entire article by Brooks is an important commentary on scientific responsibility.
26. Ad Hoc Committee on Ethical Standards in Psychological Research, *Ethical Principles in the Conduct of Research with Human Participants* (American Psychological Association, Inc., Washington, D.C., 1973; vi + 104 pp.).
27. A. De Figueiredo and J. Steele, *Manchester Guardian* (3 May 1974), p. 4.
27a. In a recent article Peter Suedfeld [*American Scientist* **63** (No. 1), 60–69 (1975)] states that the psychological damage from sensory deprivation has been greatly exaggerated in some reports, and that the technique may indeed have therapeutic value in certain conditions. We note this article here, since the

evidence it presents, based on a number of studies in different laboratories, should be examined by anyone who writes about sensory deprivation. However, it does not affect our main point, which is that some scientists and physicians have apparently played an active role in torture, and that we regard this as abhorrent in any judgment of scientific responsibility.
28. J.-P. Clavel, *New York Times* (4 Aug. 1974), section 4.
29. J. Katz, Ed., *Experimentation with Human Beings* (Russell Sage Foundation, New York, 1972; 1159 pp.).
30. "Clinical Studies in the Human: The Ethical and Scientific Problems," symposium by A. C. Barnes, R. Q. Marston, B. Barber, and J. Katz: *Fertility and Sterility* 23, 593–609 (1972).
31. B. J. Culliton, *Science* **185**, 426–427 (1974).
32. E. F. Behrman, *New York Times* (9 June 1974), Op-Ed page.
33. AAAS Committee on Science in the Promotion of Human Welfare, "The Integrity of Science," *American Scientist* 53, 174–198 (1965).
34. J. H. Comroe, Jr., *Science* **183**, 150 (1974). See also a brief account in *Environment* (St. Louis) **16** (No. 3), 21 (April 1974).
35. For a useful short history of the use of DDT, see G. McIntire, *Environment* (St. Louis) **14** (No. 6), 14 (1972).
36. T. B. Barducci, in *The Careless Technology: Ecology and International Development,* M. T. Farvar and J. P. Milton, Eds. (Natural History Press, Garden City, New York, 1972; 1030 pp.), p. 423.
37. These figures are from *WHO Relevé Epidemiologique Hebdomadaire* (8 Aug. 1969).
38. R. Garcia, "The Control of Malaria," *Environment* (St. Louis) **14** (No. 5), 2–9 (1972).
39. C. C. Hughes and J. M. Hunter, in *The Careless Technology: Ecology and International Development,* M. T. Farvar and J. P. Milton, Eds. (Natural History Press, Garden City, New York, 1972; 1030 pp.), pp. 69–93.
40. M. Kassas, *ibid.*, pp. 179–188.
41. C. J. George, *ibid.*, pp. 159–178.
42. C. Sterling, *Atlantic Monthly* (May 1974), pp. 98–105.
43. N. Wade, *Science* **185**, 234–237 (1974).
44. Of course, even in academic life, graduate students and most postdoctoral fellows work on problems largely chosen for them by the senior person with whom they have chosen to work. Most of them wish this to be so, while they are serving their apprenticeship in research. Even so, they generally have more latitude in the way they attack their problems than is possible for most workers in industrial laboratories.
45. A study by N. D. Ellis [*Technology and Society* 5, 33 (1969)] of industrial research scientists in Great Britain indicates that they do not, as a class, feel any sense of inferiority to scientists in academic life but in general accept with satisfaction their place in a different kind of scientific culture. [The major sections of Ellis's article are reprinted in Jevons (*11*), appendix 6, pp. 152–157.]
46. S. Rose and H. Rose, *Perspectives in Biology and Medicine* **16**, 605 (1973).
47. L. J. Carter and R. Gillette, *Science* **185**, 420–423 (1974); B. T. Feld, *New Scientist* **64**, 910–915 (1974).
48. P. Brodeur, *Expendable Americans* (Viking Press, New York, 1974; 274 pp.). Concerning the hazards of asbestos, and steps taken to control them, see also A.

K. Ahmed, D. R. MacLeod, and J. Carmody, *Environment* (St. Louis) **14** (No. 10), 16 (Dec. 1972).

49. Rachel Scott, *Muscle and Blood* (Dutton, New York, 1974; xiv + 306 pp.). A recent report on "New Ethical Problems Raised by Data Suppression" shows how some members of the Manufacturing Chemists Association, including scientists, had clear evidence, as early as November 1971, that vinyl chloride causes cancer in rats; yet they failed to make the facts public, or to notify the National Institute for Occupational Safety and Health, until late 1973 or early 1974. Thousands of workers were thus, for some 2 years, unnecessarily exposed to the hazards of working with vinyl chloride. The report points out: "Industrial scientists who fail to challenge conspiracies of silence within their firms are not rebuked; rather they are often quietly rewarded for their loyalty." [See *FAS* (Federation of American Scientists) *Professional Bulletin* 2 (No. 8) (November 1974).
50. Alice Hamilton, *Exploring the Dangerous Trades* (Little Brown, Boston, 1943; 433 pp.).
51. Report of the Advisory Committee on the Biological Effects of Ionizing Radiations, "The Effects on Populations of Exposure to Low Levels of Ionizing Radiation" (Division of Medical Sciences, National Academy of Sciences–National Research Council, Washington, D.C., 1972; xiv + 217 pp.).
52. No adequate history of the conflict over the SST has been written; the subject deserves the attention of social historians. The small book by W. A. Shurcliff [*SST and Sonic Boom Handbook* (Ballantine Books, New York, 1970; 154 pp.)] is naturally partisan, and was written to promote the case against the SST; but it does demonstrate a respect for the facts and contains useful references.
53. H. Johnston, *Science* 173, 517–522 (1971); *Report LBL-2217* (Lawrence Berkeley Laboratory, Berkeley, California, 1973); *Environmental Conservation* 1, 163–176 (1974).
54. For instance, by E. Hesstvedt in a lecture cited by P. Cutchis [*Science* **184**, 13–19 (1974)].
55. For references to the case of the Reserve Mining Company, see Brodeur (*48*) and W. Greene [*New York Times Magazine*, (24 Nov. 1974), p. 17].
56. Statement on "Employment Guidelines," *IEEE* (Institute of Electrical and Electronics Engineers) *Spectrum* 1973, 57–60 (April 1973).
57. S. H. Unger, "The BART Case: Ethics and the Employed Engineer" *IEEE-CSIT* (Committee on Social Implications of Technology) *Newsletter No. 4* (Sept. 1973) (Editor: Victor Klig, 497 Park Avenue, Leonia, New Jersey 07605).
58. M. S. Meselson, A. H. Westing, and J. D. Constable: Herbicide Assessment Commission of the American Association for the Advancement of Science, Background Material Relevant to Presentations at the 1970 Annual Meeting of the AAAS. Reprinted in the *Congressional Record 118* (No. 32), S3226–S3233 (3 March 1972).
59. Committee on the Effects of Herbicides in Vietnam, *The Effects of Herbicides in South Vietnam* (National Academy of Sciences, Washington, D.C., 1974; xxvi + 373 pp.).
60. Concerning this controversy, see, for instance, D. Shapley [*Science* **183**, 1177–1180 (1974)] and C. Norman [*Nature (London)* **248**, 186 (1974)].